TELEPHONE 020 7223 1144 FAX 020 7924 1112 E-MAIL dyslexia@hornsby.co.uk

PUBLISHING✓

Markco Publishing
Title: What to do when you can't learn the times tables
Author: Chinn, Steve

ISBN 1-900754-00-2

Markco Publishing
Mark College
Mark
Highbridge
Somerset
TA9 4NP

First Published in Great Britain 1996

Design and typesetting by Ibis Creative Consultants Ltd, Dulwich, London

ABOUT THIS BOOK

THE HORNSBY CENTRE
Glenshee Lodge
261 Trinity Road
London SW18 3SN
Tel: 081 874 1844
081 877 3506

TIMES TABLES! Some people, probably more than you imagine, find it very difficult to learn times tables. Their memories seem to lose the facts almost as soon as they have been learned.

This can lead to stress and to a lack of confidence with numbers.

This book offers another way to learn the times tables. It is not a short cut. It is not a magic 'cure'. It involves effort and lots of practice.

We have provided ample margins in this book. Sometimes we use them for illustration and example, but they are also for you to use for jottings, notes and practice.

When it starts to work you will find that not only are the times table facts being mastered, but that your understanding of numbers has improved too.

Number work can be made much easier with a little confidence and skill. Numbers that many people find difficult, like 9, become less threatening when connected to easier numbers, like 10. The book helps you to develop the 'easy' facts, using methods which will help other work with numbers.

When you can't remember, work it out.

Steve Chinn
February 1996

CONTENTS

TWO WAYS TO WORK

This book is organised in two sections. Each offers the same basic ideas, but in a different order.

In section 1, the times tables are taken in groups, for example, all the 2x facts are covered on pages 28 to 35.

In section 2, each times table fact is dealt with separately. This involves repeating some of the instructions, but allows you to dip into the book at places of particular need.

Remember, this is not a quick fix book. Each method will require practice and perseverance, but the child will be left with a much better understanding of numbers and a much better chance of being able to answer times tables questions.

Use what you do know to work out what you do not know.

Form a picture in your mind to help your understanding.

You may find, as the child becomes more familiar and at ease with the strategies, that she/he starts to use only a part of the strategy; enough to top up his/her memory. For example, many people are unsure if 7 x 5 is 30 or 35. Since an odd number times 5 must end in 5, the answer is 35.

THE TASK AHEAD

Times tables are often presented as individual and separate collections of facts. For example, the 2 x table is written;

$$1 \times 2 = 2$$
$$2 \times 2 = 4$$
$$3 \times 2 = 6$$
$$4 \times 2 = 8$$
$$5 \times 2 = 10$$
$$6 \times 2 = 12$$
$$7 \times 2 = 14$$
$$8 \times 2 = 16$$
$$9 \times 2 = 18$$
$$10 \times 2 = 20$$

This can be an inefficient way to learn the times tables. It does not show the connections between the facts and it presents equivalent facts as separate items eg. 4 x 3 and 3 x 4, so you have to learn the same fact twice.

e.g. 4 x 3 = 12 is in the 3x table
and 3 x 4 = 12 is in the 4x table

This book works from a TABLE SQUARE, which shows all the facts in one go.

Tables Square

	0	1	2	3	4	5	6	7	8	9	10
0	0	0	0	0	0	0	0	0	0	0	0
1	0	1	2	3	4	5	6	7	8	9	10
2	0	2	4	6	8	10	12	14	16	18	20
3	0	3	6	9	12	15	18	21	24	27	30
4	0	4	8	12	16	20	24	28	32	36	40
5	0	5	10	15	20	25	30	35	40	45	50
6	0	6	12	18	24	30	36	42	48	54	60
7	0	7	14	21	28	35	42	49	56	63	70
8	0	8	16	24	32	40	48	56	64	72	80
9	0	9	18	27	36	45	54	63	72	81	90
10	0	10	20	30	40	50	60	70	80	90	100

There are advantages in using the square. These will become clear as you progress through the book.

There are four blank table squares and one extra completed one printed at the back of the book. The blank squares are for keeping a record of progress. The child can fill in each section as he or she learns the new facts.

The completed square is the target.

So you cannot wait. You want to use the tables square now.

So what is 5 x 7?

Find 5 along the top row. Then run your finger down that column until it is in the line which has 7 on the left hand side.

The answer is 35.

HOW TO HALVE THE TASK

121
to go

There are 121 facts to learn, but

that can be reduced to almost a half by demonstrating

one mathematical idea:

the order in which you multiply two numbers

together does not matter, the answer will be the same

If 4 x 7 = 28

What is 7 x 4 = ?

So   **6 x 4 = 24**   is the same as   **4 x 6 = 24**

and   **7 x 9 = 63**   is the same as   **9 x 7 = 63**

and so on.

If 9 x 10 = 90

What is 10 x 9 ?

6 x 4, 7 x 9 and other examples can be demonstrated

with rows and columns of 1p coins. For example, 24

coins can be set up as shown in the diagram opposite.

This can be viewed as 6 columns of 4 (6 x 4) or as 4

rows of 6 (4 x 6). It is obvious that the answer (24) is

the same for both.

In general

number A x number B   =   number B x number A

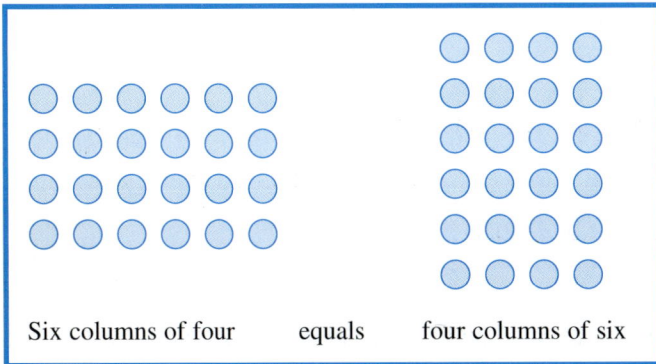

Six columns of four   equals   four columns of six

If you look at the table square on page 7, you will see that the numbers in the left diagonal are repeated in the right diagonal. If you folded the square diagonally so that the two stars were brought together, the numbers in the two diagonals are the same.

There are really only 66 different facts to learn.

WORDS USED FOR X

Let's look at some of the words which can mean **X**

multiply   **times**

of   **product**

lots of

and sometimes nothing at all as in 'six sevens' or 'three nines'.

$7 \times 4 = 28$

$10 \times 9 = 90$

Examples:

Three multiplied by four is twelve.

Three times four is twelve.

The product of three and four is twelve.

Three fours are twelve.

WHAT IS MULTIPLICATION

Multiplication is just a quick way of adding together the same number several times. This is repeated addition, the same number added to itself several times. You can explain how the different words and phrases for multiplication help the understanding of the idea and its use. The "lots of" is particularly useful to show the link between multiplication and addition.

For example,

4 x 6   is also   6 + 6 + 6 + 6   (4 lots of 6)

   or

7 x 5   is also   5 + 5 + 5 + 5 + 5 + 5 + 5   (7 lots of 5)

GETTING A PICTURE

For many children the times table facts have no reality. They are just meaningless, unconnected abstract symbols. Learning is much more effective if meaning and image can be added. In mathematics, teachers sometimes use simple equipment to demonstrate an idea. This book suggests you use coins, number strips and squares.

There are some squares and number strips supplied with this book. They are to help your child to learn. If you use them for a while and you do not find them helpful, then stop using them.

You could also use coins to help form a stronger picture of the times table facts.

Let's start with the squares.

Make a line of 4 squares

That is 1 'lot of' 4 (1 x 4).

Make a second line of 4 squares and put them together with the first line.

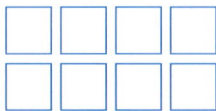

That makes 2 'lots of' 4 (2 x 4).

The area you have made also represents 2 'lots of' 4 or 2 x 4 (=8).

I had a bar of chocolate with 18 portions, which I wished to share between six people. How many portions would each person receive if everybody is to have the same amount?

If I wished to share between three people, how many portions would each person receive if everybody is to have the same amount?

If I wanted to give three people six pieces of chocolate how many pieces would I need?

If I wanted to give six people three pieces of chocolate how many pieces of chocolate do I need?

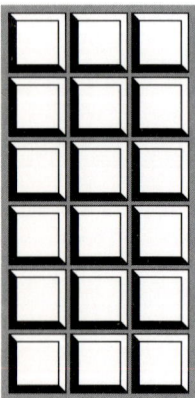

It is also clear that this area can be considered either as 4 vertical columns of 2, that is 4 x 2

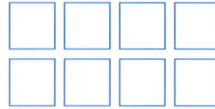

or 2 horizontal rows of 4, that is 2 x 4.

This and similar examples confirm the idea we met on page 5.

2 x 4 = 4 x 2.

number A x number B = number B x number A

The use of the squares in this way gives two pictures or ideas for times......... **lots of** and **an area**. You could use these pictures and ideas to help the child's memory as you work through this book.

You will also find coins are used to illustrate or give a picture of some strategies.

STARTING

When the task ahead of you seems very difficult it can be very hard to get started. What you will be able to do now is to demonstrate how quick progress can be encouraging. This is additional encouragement to the fact that number A x number B is the same as number B x number A, which effectively halved the task.

SOME QUICK PROGRESS

(Note that zero, 0 and one, 1 are a little exceptional, and do not set a pattern for the other numbers).

Think of multiplication as producing a shape, an area. Three times five is three columns by five rows. So we see an area. We could even count the number of square or coins.

The biggest number we meet in this book is 100 (10 x 10). Try to relate other areas to this key area. So, for example, 5 x 5 is 25, an area that is one quarter of the 100. 7 x 7 is 49, an area that is almost one half of the 100.

0

0 Nil

ZERO

Nothing

0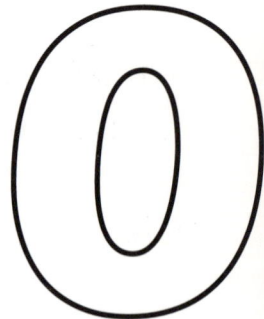

None

ZERO   0

The answers to the zero times table are easy to remember, because THEY ARE ALL ZERO.

When you multiply any number by zero, the answer is always zero.

For example;

$$5 \times 0 = 0 \qquad 1\ 000\ 000 \times 0 = 0$$

If you use the 'lots of' for multiply, then you can see that this result is obvious.

5×0 is 5 lots of zero (zero also means none or nothing) which must (still) be zero.

$$5 \times 0 = 0.$$

$1\ 000\ 000$ is one million lots of zero which still is zero.

$$1\ 000\ 000 \times 0 = 0$$

Even a million lots of nothing is still nothing!

The same applies when you have

zero   times   a number.

0   x   number

Did you know...
.... that the Romans had no zero, because they could not see the point of having a symbol for nothing. How can you represent nothing with something. So they ignored it!

Quick Quiz

a) What is 0 x 0 x 0 = ?

b) What is the sum of 0 + 0 ?

c) If there are 15 girls in a class and 17 boys, and each of the girls has no apples and each of the boys has no pears, how many pieces of fruit do they have altogether?

For example;

$$0 \times 5 = 0$$

$$0 \times 1\ 000\ 000 = 0$$

If you use 'lots of' for 'times' then these examples say;

Zero lots of 5 is zero $0 \times 5 = 0$

Zero lots of one million is zero $0 \times 1\ 000\ 000 = 0$.

Zero lots of huge numbers such as millions is still nothing!

This is another example of

number A x number B = number B x number A

$$1\ 000\ 000 \times 0 = 0 \times 1\ 000\ 000 = 0$$

YOU HAVE LEARNED 21 FACTS ON

THE TABLE SQUARE.......100 TO GO!

100 to go

1 One

Unity

Unity

 *Single*

1p

Unique

1

ONE   1

The answers to the 1 x table are the same value as the

multiplying number.

One 'lots of' any number must be the same as that

number.

For example; 1 'lot of' 6 equals 6

$$1 \times 6 = 6$$

The answers to the 1 x table are the same as the number

being multiplied by 1.

$$1 \times 1 \ = 1$$
$$1 \times 2 \ = 2$$
$$1 \times 3 \ = 3$$
$$1 \times 4 \ = 4$$
$$1 \times 5 \ = 5$$
$$1 \times 6 \ = 6$$
$$1 \times 7 \ = 7$$
$$1 \times 8 \ = 8$$
$$1 \times 9 \ = 9$$
$$1 \times 10 = 10$$

Again the other order of multiplying the two numbers

(number x 1) must give the same answer, so that

$$1 \times 9 = 9 \times 1.$$

This is clear if 'lots of' is again used for x;

9 'lots of' 1 equals 9

9 x 1 = 9 (which is the same as 1 x 9 = 9)

 1 x 1 = **1**

 2 x 1 = **2**

 3 x 1 = **3**

 4 x 1 = **4**

 5 x 1 = **5**

 6 x 1 = **6**

 7 x 1 = **7**

 8 x 1 = **8**

 9 x 1 = **9**

 10x 1 = **10**

Both of these ideas can be demonstrated by using coins.

For example, 6 lots of 1p   6 x 1p = 6p

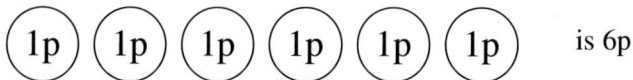   is 6p

and 1 lot of 6 coins is also 6p

Another example uses a 5p coin to show that 1 lot of 5p

is the same as 5 lots of 1p.

THIS GIVES YOU 19 NEW FACTS ON

THE TABLE SQUARE.....81 TO GO!

81
to go

10

TEN

10 = 2 x 5

decade

10p

10

decimal

decimetre

10 20 30 40 50 60 70 80 90 100

TEN   10

The answers to the 10x table are the same as the

multiplying numbers, but have a zero in the units place

(making them 10 times bigger).

10 x 1 = 10

10 x 2 = 20

10 x 3 = 30

10 x 4 = 40

10 x 5 = 50

10 x 6 = 60

10 x 7 = 70

10 x 8 = 80

10 x 9 = 90

10 x 10 = 100

Ten is an important and very useful number. We organise our number system in units, tens, hundreds, thousands, ten thousands, hundred thousands, millions and so on.

This is because we have a calculator which gives us ten units. It is our hands!

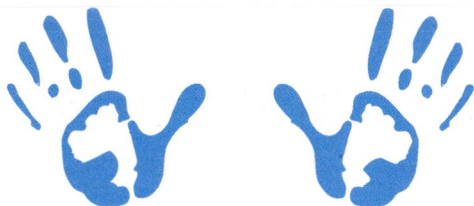

Quick fact

The latin for ten is *decem.* The number system based on ten units is called a decimal system.

The latin for 100 is *centum,* which gave us century (which is 10 x 10).

Quick facts

If one dot is 1 millimetre across (that is about the size of a pin head) -

Then if you joined dots end to end
10 dots cover 1 cm
100 dots are 10 cm
1 000 dots are 1 metre
10 000 dots are ten metre
100 000 dots are 100 metres
and 1 000 000 - that one million dots will be one kilometre!

If we were using squares
100 dots cover an area
1 cm x 1 cm

1 000 000 dots are one metre square

So if ten pixies can dance on a pin head just one millimetre across, and a football pitch is 100 metres square, how many dancing pixies can dance in the area of a football pitch?

All these numbers are based on "lots of" 10.

10	Ten is 10 units.
100	A hundred is 10 tens.
1 000	A thousand is 10 hundreds.
10 000	Ten thousand is 10 thousands.
100 000	A hundred thousand is 10 ten thousands.
1 000 000	A million is 10 hundred thousands.

Numbers often group in patterns. There is an easy pattern for the numbers above. The next number below each time has one more zero, one more digit.

So ten,10, is a two digit number,

a one and one zero

one hundred,100, is a three digit number,

a one and two zeros

one thousand, 1000, is a four digit number,

a one and three zeros

ten thousand, 10 000, is a five digit number,

a one and four zeros

one hundred thousand, 100 000, is a six digit number,

a one and five zeros

one million, 1000 000, is a seven digit number.

a one and six zeros

10 is the first two digit number. It is made from 10 units.

Counting in tens develops so that;

2 lots of 10 are written as 20, (2 lots of 10 units)

3 lots of 10 are written as 30, (3 lots of 10 units)

4 lots of 10 are written as 40, (4 lots of 10 units)

and so on.

These examples could be illustrated with 1p and 10p coins. As with all illustrations, it helps if the child handles the coins.

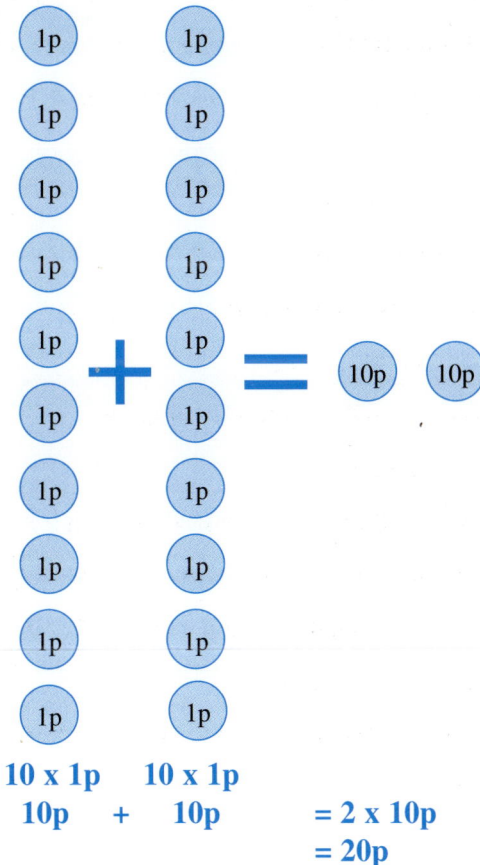

10 x 1p   10 x 1p
10p   +   10p   = 2 x 10p
**   = 20p**

A hundred billion - which is 100,000,000,000!

23

Now compare the values of 1's and 10's. Ask the child to set up rows of 1p and 10p coins next to each other, with the same number of coins in each row. Discuss the value of each group of coins.

For example, the 4 x 1p coins show 4p.

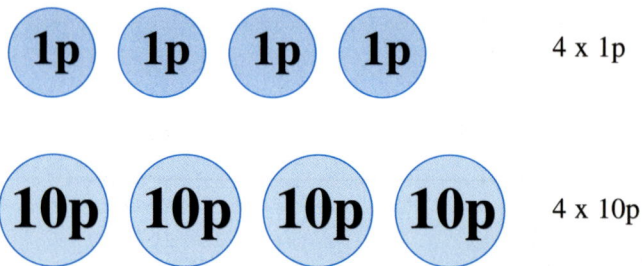

(1p) (1p) (1p) (1p)   4 x 1p

(10p) (10p) (10p) (10p)   4 x 10p

and the four 10p coins row shows 40p.

The 10p row shows 4 coins, each worth 10 times a 1p coin. The value of 40p in the 10p row is 10x that of the 4p in the 1p row.

Another example could be made using the 10 strips and 1 strips from the back of the book. 4 of the 10 strips will make 40 a line 10 times as long as 4 of the 1 strips.

If you change over from 'lots of' to 'times' or 'x', then
you have the ten times table.

1 x 10 = **10**	10 x 1 = **10**
2 x 10 = **20**	10 x 2 = **20**
3 x 10 = **30**	10 x 3 = **30**
4 x 10 = **40**	10 x 4 = **40**
5 x 10 = **50**	10 x 5 = **50**
6 x 10 = **60**	10 x 6 = **60**
7 x 10 = **70**	10 x 7 = **70**
8 x 10 = **80**	10 x 8 = **80**
9 x 10 = **90**	10 x 9 = **90**
10 x 10 = **100**	10 x10 = **100**

There is also a pattern in the sounds of this times table;

one times ten is ten

two times ten is twenty (two-ty)

three times ten is thirty (three-ty)

four times ten is forty

five times ten is fifty (five-ty)

six times ten is sixty

seven times ten is seventy

eight times ten is eighty

nine times ten is ninety

ten times ten is one hundred (not such a good sound
match!)

THIS GIVES YOU 17 NEW FACTS....64 TO GO (AND DON'T FORGET, YOU CAN REDUCE THAT TO ALMOST HALF, 36, SINCE A x B = B x A).

THE 2X AND 10X FACTS CAN BE USED TO WORK OUT ALL THE OTHER FACTS. THEREFORE IT IS VERY IMPORTANT FOR THE CHILD TO DEVELOP RAPID RECALL OF THESE FACTS.

A KEY SKILL FOR THE NEXT WORK IS BREAKING DOWN AND BUILDING UP NUMBERS. IF THESE CAN BE RECALLED AUTOMATICALLY THEN THE CHILD CAN BUILD ALL THE OTHER FACTS FROM THEM.

Coins are a very good material to use to demonstrate number values. Most children are familiar with coins and their values (not all children are, so this work will have extra benefits for them).

Have you noticed how coins are only available in certain values? Then values are based on multiples of 1, 2 and 5.

We have a 1p   2p   5p   10p   20p   50p   100p (£1), and soon 200p (£2) and 500p (£5)

We combine these to make other values, for example;

3p can be made from   2p + 1p

6p can be made from   5p + 1p

7p can be made from   5p + 2p

4p can be made from   2 lots of 2p   (2 x 2p)

One way you could give someone 9p is to give them 10p and ask them to give you 1p back.

So   9p can be made from   10p - 1p

This means that 3, 4, 6, 7 and 9 can all be made up from 'easier' numbers.

$3 = 2 + 1$

$4 = 2 \times 2$

$6 = 5 + 1$

$7 = 5 + 2$

$9 = 10 - 1$

5 can also be related to 'easy' 10 because

5 is half of 10 $5 = 10 \div 2$

THIS IS GIVES YOU A STRATEGY TO WORK
OUT ALL THE OTHER FACTS USING JUST
THE 2x and 10x FACTS.

2

Pair

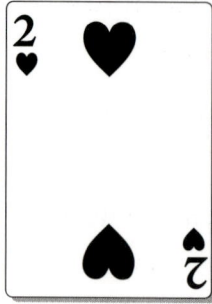

twice

twin

bi-

GEMINI

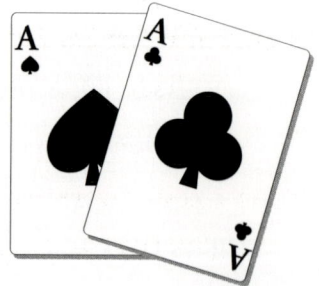

DUO

2p

TWO 2

THIS IS A VERY IMPORTANT TABLE TO LEARN,
BECAUSE YOU WILL USE IT TO WORK OUT
MANY OTHER TIMES TABLE FACTS.

*The pattern for 2x is the pattern for even numbers (even
numbers are those numbers which divide equally into
two parts).*

The first five even numbers set the pattern

     2   4   6   8   10

Other even numbers use the same five digits, that is, a
2 or 4 or 6 or 8 or 0 in the unit place.

The next five even numbers show this;

    12   14   16   18   20

as do the next five even numbers

    22   24   26   28   30

A lot of children who think they 'know' the 2 times table can only work out separate answers (eg 7 x 2) if they start at the beginning (1 x 2) and count up in 2's. You can show a child how to short circuit this a little, and teach a useful strategy at the same time, by breaking down the numbers as shown on page 26.

The strategy makes use of 5 x 2 = 10 as a half way stage, instead of going back to the beginning, 1 x 2.

So 5 x 2 = 10 is a useful half way point.

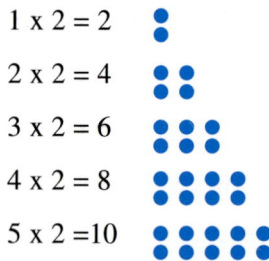

1 x 2 = 2

2 x 2 = 4

3 x 2 = 6

4 x 2 = 8

5 x 2 =10

(You could encourage the child to think of 5 in terms of the fingers on one hand, that is, half of his/her ten fingers)

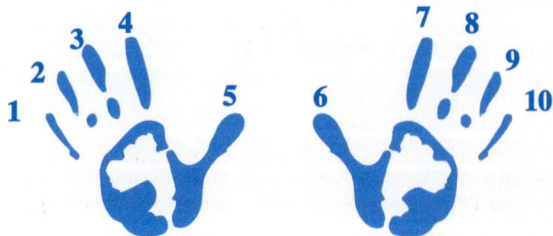

This strategy, then, uses the easy fact that 5 lots of 2 equal 10   (5 x 2 = 10)

This is like taking 5 lots of 2p coins and trading them for one 10p coin. It is useful for the child to handle and arrange the coins that show this strategy.

(2p) (2p) (2p) (2p) (2p) = (10p)

The strategy then builds on the addition facts that relate 6, 7, 8 and 9 to 5.

6 = 5 + 1   7 = 5 + 2   8 = 5 + 3   9 = 5 + 4

(or 9 = 10 - 1)

6 lots of 2 (6 x 2) can be worked out as

5 lots of 2   plus 1 more lot of 2;

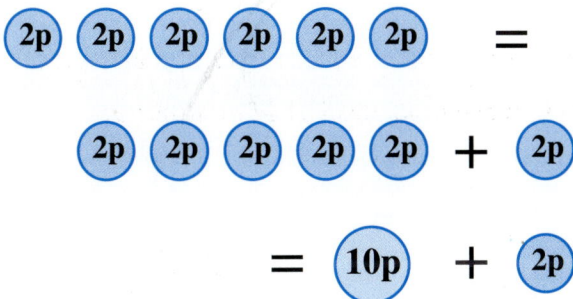

(2p) (2p) (2p) (2p) (2p) (2p)   =

(2p) (2p) (2p) (2p) (2p) + (2p)

= (10p) + (2p)

6 x 2   =   5 x 2   + 1 x 2

6 x 2   =   10   + 2 = 12

Encourage the child to use 2p and 10p coins to work this (and other examples) through.

7 lots of 2, (7 x 2) can be worked out as *5 lots of 2* plus *2 more lots of 2, which can be shown using 2p coins.*

$$= \qquad\qquad +$$

7 x 2	=	5 x 2	+ 2 x 2		
7 x 2	=	10	+ 4	=	14

8 lots of 2, (8 x 2) can be worked out as *5 lots of 2* plus *3 more lots of 2.*

○ ○ ○ ○ ○ ○ ○ ○ = ○ ○ ○ ○ ○ + ○ ○ ○

8 x 2	=	5 x 2	+ 3 x 2		
8 x 2	=	10	+ 6	=	16

9 lots of 2, (9 x 2) can be worked out as *5 lots of 2* plus *4 more lots of 2*

○ ○ ○ ○ ○ ○ ○ ○ ○ = ○ ○ ○ ○ ○ + ○ ○ ○ ○

8 x 2	=	5 x 2	+ 3 x 2		
8 x 2	=	10	+ 6	=	16

The strategy is to start at 5 x 2, which has an easy answer of 10 and add on the other bits you need to make 6 x 2, 7 x 2, 8 x 2, 9 x 2.

You can also work out 9 x 2 by using 9 = 10 - 1

9 x 2 is 9 lots of 2

    which equals *10 lots of 2*   minus   *1 lot of 2*

                     =   -

9 x 2	=	10 x 2	- 1 x 2
9 x 2	=	20	- 2 = 18

We use this break down and build up strategy a lot in this book.

Now practise!

For example, when trying 7 x 2, ask the child to use the two stages, saying them out loud to help input to the memory:

"7 times 2 is 5 times 2,   **7 x 2 = 5 x 2,**

    that's 10, plus 2 times 2,   **that's 10 + (2 x 2)**

       that's 4, that's 14."   **that's 4, that's 14.**

Try to build a picture of 7 as 5 + 2 in the child's mind.

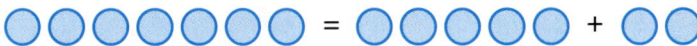

○○○○○○○ = ○○○○○ + ○○

You could use a 5p coin and a 2p coin, for example or fingers. Try to notice, or, better still, ask which the child prefers for his/her picture / image.

Also again try letting the child make the sum with coins as he/she says the sum, setting out 5 lots of 2p, then trading them for a 10p, and 2 lots of 2p, making 14p altogether.

Reminders

For the first 2 times facts, it can help to remember the old rhyme

"Two, four, six, eight,

Who do we appreciate?"

$$1 \times 2 = 2$$
$$2 \times 2 = 4$$
$$3 \times 2 = 6$$
$$4 \times 2 = 8$$

This sets the pattern. The facts after 5 x 2 follow the same pattern, but with a ten's digit in front of each 2, 4, 6, 8.

So we can see the pattern

1 x 2 = 2	6 x 2 = 12
2 x 2 = 4	7 x 2 = 14
3 x 2 = 6	8 x 2 = 16
4 x 2 = 8	9 x 2 = 18

Breaking down 6, 7, 8 and 9 into 5 + 1, 5 + 2, 5 + 3 and 5 + 4 is just the same as if you were counting on your fingers (thumbs act as fingers for counting purposes). 5 is the number of fingers on one hand, and the +1, +2, +3 and +4 are the numbers of fingers on your other hand.

YOU NOW HAVE A FURTHER 15 FACTS TO ADD TO YOUR BLANK TABLES CHART. ONLY 49 TO GO! (and only 28 different facts).

49
to go

5

$$10 \div 2 = 5$$

Pentagon

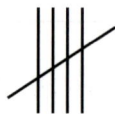

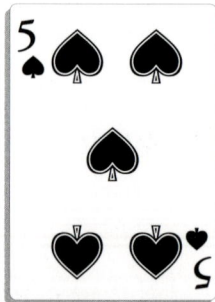

V

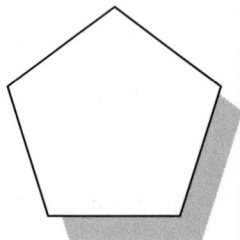

5p

Quins

5 10 15 20 25 30 35 40 45 50

FIVE   5

Five is half of ten. The 5x table is worked out by halving the 10x table facts.

$$5 = 10 \div 2 \qquad \text{or} \qquad \frac{10}{2}$$

The 5 times table can be worked out quickly by halving the 10 times table, that is dividing the answer to the equivalent 10x fact by 2.

Try to establish the idea. Take some 5p and 10p coins and work through some examples like those below with the child.

Set up next to each other 3 of the 5p coins and 3 of the 10p coins.

3 x 5p = 15p

3 x 10p = 30p

The three 10p coins are worth 30p and the three 5p coins are worth half as much, 15p.

Try again with 8 of the 5p coins and 8 of the 10p coins.

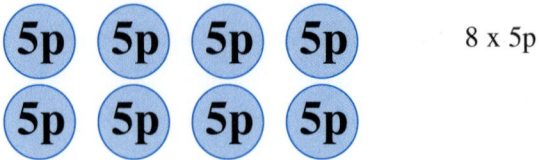

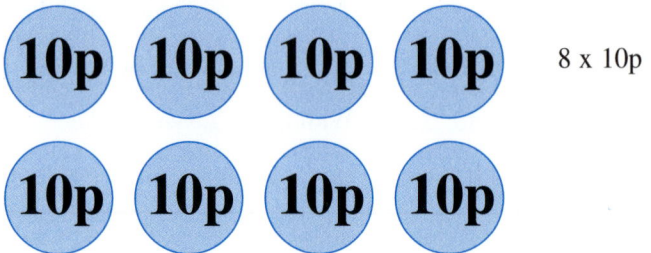

5p 5p 5p 5p   8 x 5p
5p 5p 5p 5p

10p 10p 10p 10p   8 x 10p

10p 10p 10p 10p

The 8 x 10p coins are worth 80p and the 8 x 5p coins are worth half as much, 40p.

You can see that in each example, a number of 5p coins is worth half the value of the same number of 10p coins.

Look at the 5 x table and, next to it in blue, the answers for the 10x table. Discuss with the child how the 5x answers are all half as much as the 10x answers. See in the first three, how 5 is half of 10, 10 is half of 20, and 15 is half of 30.

$$1 \times 5 = \quad 5 \qquad 10$$

$$2 \times 5 = \quad 10 \qquad 20$$

$$3 \times 5 = \quad 15 \qquad 30$$

Now look at the whole 5x table

1 x 5 =	5	10
2 x 5 =	10	20
3 x 5 =	15	30
4 x 5 =	20	40
5 x 5 =	25	50
6 x 5 =	30	60
7 x 5 =	35	70
8 x 5 =	40	80
9 x 5 =	45	90
10 x 5 =	50	100

Each 5x answer is half the value of the 10x answer (in blue) next to it.

Also, use the simple pattern in the 5x table to give an additional check. Look at the units digits as you go down the 5x table. They alternate:

5 - 0 - 5 - 0 - 5 - 0 - 5 - 0 - 5 - 0

This pattern links to the number which is multiplying the 5.

If it is an even number times 5 (2x5, 4x5, 6x5, 8x5, 10x5), then the answer has 0 as its unit digit.

If it is an odd number times 5 (1x5, 3x5, 5x5, 7x5, 9x5), then the answer has 5 as its unit digit.

Times Check

For 5x
If it's even it ends in 0
If it's odd it ends in 5

2 x 5 =   **10**

4 x 5 =   **20**

6 x 5 =   **30**

8 x 5 =   **40**

10 x 5 =   **50**

1 x 5 =   **5**

3 x 5 =   **15**

4 x 5 =   **25**

7 x 5 =   **35**

9 x 5 =   **45**

36
to go

So, to work out a 5x answer, start with a 10x answer, halve it and check back to see if it was an odd number of 5's (then the answer will end in 5) or an even number of 5's (when the answer will end in 0).

Let us try some examples.

7 x 5

Step 1:   7 x 10 = 70

Step 2:   70 ÷ 2 = 35

Step 3:   Check, 7 is an odd number, 35 ends in 5.

6 x 5

Step 1:   6 x 10 = 60

Step 2:   60 ÷ 2 = 30

Step 3:   Check, 6 is an even number, 30 ends in 0.

Now practise!

For each example used ask the child to quote the 10x answer before the 5x answer and to predict if the units digit will be a 5 or a 0.

THIS GIVES 13 NEW FACTS, REDUCING THE NUMBER TO LEARN TO 36 (of which only 21 are different).

Quadruple

4

Quadrilateral   **2 x 2**

IV

Quads

4   8   12   16   20   24   28   32   36   40

4

Games to play

Try this little game.

Think of a number
Double it
Double it again
Add 4
Divide by two
Divide by two again
Take away the number you
first thought of.
The answer is 1!

You should be able to work
out how this works.

If instead of saying "add 4"
you say "add 8", then the
answer is 2. Indeed
whatever you add, if you
divide it by 4, then that is
the answer.

FOUR   4

*To multiply a number by 4, you double it, then double it
again. That is, you multiply by 2 twice.*

You have just seen how to work out 5x facts by dividing
the 10x table by 2.

You can show the child how to work out the 4x table by
multiplying the 2x table facts by 2 again.

This works because 4 is 2 x 2.

Look at the 2x and the 4x tables written next to each
other. The answers to the 4x table are twice the answers
to the 2x table.

1 x 2 = 2	1 x 4 = 4
2 x 2 = 4	2 x 4 = 8
3 x 2 = 6	3 x 4 = 12
4 x 2 = 8	4 x 4 = 16
5 x 2 = 10	5 x 4 = 20
6 x 2 = 12	6 x 4 = 24
7 x 2 = 14	7 x 4 = 28
8 x 2 = 16	8 x 4 = 32
9 x 2 = 18	9 x 4 = 36
10 x 2 = 20	10 x 4 = 40

To work out individual 4x facts, the child has to remember to multiply by 2 twice.

For example,

to work out 4 x 7;

Use two easy steps: 2 x 7 = 14;   2 x 14 = 28.

to work out 4 x 3;

Use two easy steps: 2 x 3 = 6;   2 x 6 = 12.

When you ask the child to practise, remember to encourage her/him to talk through both steps.

THIS GIVES YOU 11 NEW FACTS.

ONLY 25 TO GO.

25
to go

9

9

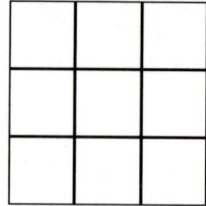

3 x 3

IX

9 18 27 36 45 54 63 72 81 90

NINE   9

The 9x facts are calculated by using a simple process using the 10x table and an easy subtraction or addition. Checking is by an easy addition.

This method introduces the useful skill of ESTIMATION. The method is based on the fact that 9 is close in value to 10, that it is 1 less than 10.

Nine is the number before ten. Ten is an easy number to work with, so you show the child how to work with 10 instead of 9 (this is an estimate) and then readjust the answer back to be the correct 9x fact.

This estimation skill will help the child in other areas of mathematics, too. For example, many items in shops are priced to be £1.99, £49.99, £6.95 and so on.

These are much easier to add if you adjust them to £2, £50 and £7 and then readjust by taking off the 7p (that is 1p + 1p + 5p) you added to round up the prices, from the total of £59 to give £58.93.

Start by showing money examples to give the idea of how 9 values relate closely to 10 values.

Then develop the pattern for the 9x table.

A good image to help understanding uses the 1 unit, 9 unit and 10 unit strips at the back of the book. Cut them out carefully.

Set up two lines of the paper strips. One will be "lots of" 10 and the other, placed next to it, will be "lots of" 9. Then even up the lines by using "lots of" 1 to make the two lines of equal length.

First put down 1 of the 10 strips.

Next to it put 1 of the 9 strips.

Put 1 of the 1 strips at the end of the nine strip to make it equal in length to the 10 strip.

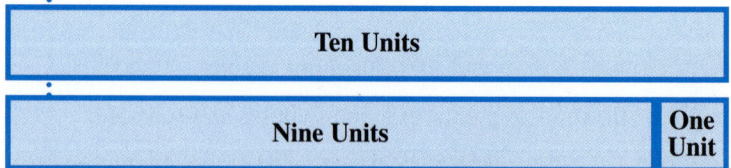

This shows the first step in the pattern

1 nine is 1 less than 1 ten.

1 x 9 = 1 x 10 - 1 x 1

Now make the line longer.

Add a second 10 strip to the 10 line.

Add a second nine strip to the 9 line.

You will need a second 1 strip to make the two lines equal again.

2 nines are 2 less than 2 tens

2 x 9 = 2 x 10 - 2 x 1

Now make the line longer again.

Add a third 10 strip to the 10 line.

Add a third 9 strip to the 9 line.

3 nines are 3 less than 3 tens

3 x 9 = 3 x 10 - 3 x 1

The pattern is there, and you could now continue for any number of nines compared to the same number of tens, for example;

7 nines are 7 less than 7 tens

7 x 9 = 7 x 10 - 7 x 1

Now look together at the 10x table alongside the 9x table, with the number of 1's needed to adjust the 10x table into the 9x table.

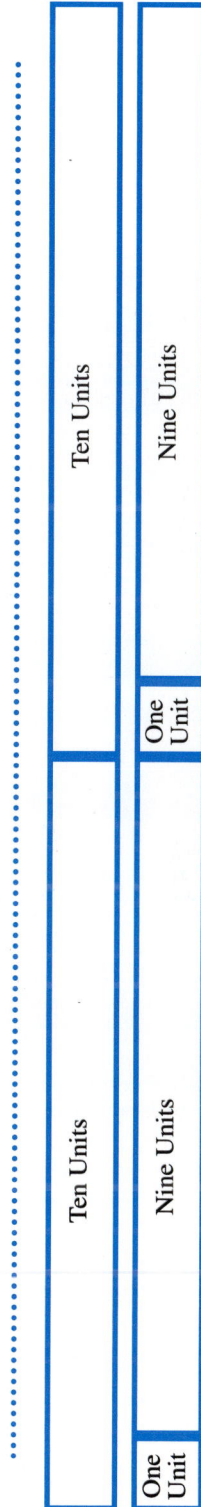

Ten Units

Nine Units

One Unit

Ten Units

Nine Units

One Unit

$1 \times 9 = 9$ ($1 \times 10 - 1 \times 1$ $= 10 - 1 = 9$)

$2 \times 9 = 18$ ($2 \times 10 - 2 \times 1$ $= 20 - 2 = 18$)

$3 \times 9 = 27$ ($3 \times 10 - 3 \times 1$ $= 30 - 3 = 27$)

$4 \times 9 = 36$ ($4 \times 10 - 4 \times 1$ $= 40 - 4 = 36$)

$5 \times 9 = 45$ ($5 \times 10 - 5 \times 1$ $= 50 - 5 = 45$)

$6 \times 9 = 54$ ($6 \times 10 - 6 \times 1$ $= 60 - 6 = 54$)

$7 \times 9 = 63$ ($7 \times 10 - 7 \times 1$ $= 70 - 7 = 63$)

$8 \times 9 = 72$ ($8 \times 10 - 8 \times 1$ $= 80 - 8 = 72$)

$9 \times 9 = 81$ ($9 \times 10 - 9 \times 1$ $= 90 - 9 = 81$)

$10 \times 9 = 90$

So, for example, to work out 6 x 9, you start with 6 x 10 = 60 and then subtract 6, which gives 54.

This will work for ANY number times 9.

For example, 34 x 9 can be done by starting with 34 x 10 = 340 and then subtracting 34, which gives 306.

More patterns.

Look at the answers to the 9x table. If you add up the tens and units digits for each answer you ALWAYS get 9.

Examples; $7 \times 9 = 63$, $6 + 3 = 9$

                   $2 \times 9 = 18$, $1 + 8 = 9$

ALL answers to ANY 9x sum follow this rule.

For example 888 x 9 = 7992.

7 + 9 + 9 + 2 = 27,

then 2 + 7 = 9.

This is a useful check, and can also help working out an answer to the 9x table. For example, 7 x 9 can be worked through by saying "7 x 9 is less than 7 x 10 and will be sixty something, where the something is the units digit. The units digit is the 3 you add on to 6 to make 9.

THIS GIVES 9 NEW FACTS.

THERE ARE ONLY 16 TO GO.

NEARLY THERE!

Quick calculation!

Here is an odd little trick you can do with a calculator.

Put in the number 12345679
(careful, there is no 8)

Then multiply it by any multiple of 9.

So for example try 12345679 x 27 =

See the pattern?

16
to go

3 6 7 8

Triplicate
Triplicate
Triplicate

Octopus

MONDAY
TUESDAY
WEDNESDAY
THURSDAY
FRIDAY
SATURDAY
SUNDAY

Sextant

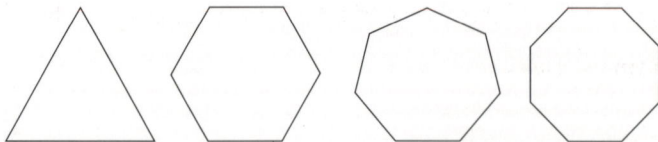

THREE 3, SIX 6, and SEVEN 7

3x, 6x and 7x are considered as 3 'lots of', 6 'lots of' and 7 'lots of'. Each of these is then broken down..

   3 'lots of' is '2 lots of' plus 1 'lots of'.

   6 'lots of' is 5 'lots of' plus 1 'lots of' and

   7 'lots of' is 5 'lots of' plus 2 'lots of'.

These three sets of facts can be worked out by effectively the same strategy. The harder fact, whether it is 3x , 6x , 7x , is broken down into two easier facts, using the 'lots of' interpretation of times. This strategy will help mop up most of the remaining facts.

One of the strategies which has been used regularly in this book is the strategy of breaking down and building up numbers. This strategy is used a great deal in mathematics, for example when multiplying bigger numbers together, such as 64 x 35. So this skill will have further use after the times tables are mastered.

We are going to tackle the last few remaining facts from two directions. This means, for example, we will work out 3 x 7 and then work out 7 x 3 by a variation of the same strategy.

Breaking down to build up

When we do the sum
   64 x 35 = ?
we start by doing the sum
   64 x 3 (= 192)
we then multiply the answer by 10, which we know is just work in a '0'.
   192 x 10 = 1920
Then we do the sum
   64 x 5 (=320)
And finally we add the two together.
   1920 + 320 = 2240

Direction 1

We are going to break down

  6 times into (5 times + 1 times)

and

  7 times into (5 times + 2 times)

and

  3 times into (2 times + 1 times)

Or to put it another way, we are going to use

5 'lots of' + 1 'lot of' to work out 6 'lots of'

and

5 'lots of' + 2 'lots of' to work out 7 'lots of'

and

2 'lots of' + 1 'lot of' to work out 3 'lots of'.

This means using; (5x plus 1x) for 6x

         (5x plus 2x) for 7x

         (2x plus 1x) for 3x

So for the remaining 3x facts;

(you already know 3 x 2, 3 x 4, 3 x 5, 3 x 9 and 3 x 10)

3 x 3 = 2 lots of 3 plus 1 lot of 3

     = 3 + 3   +   3

     = (2 x 3)   +   (1 x 3)

     = 6   + 3 = 9

3 x 3 **= 9**

In the same way we can look at the others:-

3 x 6   =   2 lots of 6   plus   1 lot of 6

      =   6 + 6   +   6

      =   (2 x 6)   +   (1 x 6)

      =   12   +   6

      =   18

3 x 6   =   **18**

3 x 7   =   2 lots of 7   plus   1 lot of 7

      =   7 + 7   +   7

      =   (2 x 7)   +   (1 x 7)

      =   14   +   7

      =   21

3 x 7   =   **21**

3 x 8   =   2 lots of 8   plus   1 lot of 8

      =   8 + 8   +   8

      =   (2 x 8)   +   (1 x 8)

      =   16   +   8

      =   24

3 x 8   =   **24**

For the remaining 6 x facts; (again, you know 6 x 2, 6 x 3, 6 x 4, 6 x 5, 6 x9 and 6 x 10)

6 x 6 = 5 lots of 6   plus   1 lot of 6

       = 6+6+6+6+6   +   6

       = (5 x 6)   +   (1 x 6)

       = 30   +   6

       = 36

6 x 6 = **36**

6 x 7 = 5 lots of 7   plus   1 lot of 7

       = 7+7+7+7+7   +   7

       = (5 x 7)   +   (1 x 7)

       = 35   +   7

       = 42

6 x 7 = **42**

6 x 8 = 5 lots of 8   plus   1 lot of 8

       = 8+8+8+8+8   +   8

       = (5 x 8)   +   (1 x 8)

       = 40   +   8

       = 48

6 x 8 = **48**

Now for the remaining 7x facts: (you already know, 7 x 2, 7 x 3, 7 x 4, 7 x 5, 7 x 6 and 7 x 9)

7 x 7 = 5 lots of 7   plus   2 lots of 7

       = 7+7+7+7+7   +   7+7

       = (5 x 7)   +   (2 x 7)

       =   35   +   14

       =   49

7 x 7 = 49

7 x 8 = 5 lots of 8   plus   2 lots of 8

       = 8+8+8+8+8   +   8+8

       = (5 x 8)   +   (2 x 8)

       =   40   +   16

       =   56

7 x 8 = 56

All that is left is 8 x 8.

And 8 x 8 is rather special. Each 8 can be made by multiplying 2 x 2 x 2, so 8 x 8 is the same as multiplying 2 x 2 x 2 x 2 x 2 x 2, and the answer goes through a useful sequence, starting with 2, then

2 x 2 = 4,   2 x 4 = 8,   2 x 8 = 16,

2 x 16 = 32,   2 x 32 = 64.

General Practice

Talk the child through examples, such as:

3 x 8   is   (2 x 8)   +   (1 x 8)

  which is (16 + 8)

    which is 24   (add 8 to 16 by adding it as 4 then 4)

6 x 4   is   (5 x 4)   +   (1 x 4)

  which is (20 + 4)

    which is 24

7 x 8   is   (5 x 8)   +   (2 x 8)

  which is (40 + 16)

    which is 56

And many more examples from the Tables Square.

Direction 2

With Direction 1, you broke down the multiplying number, the first number in the times sum.

With Direction 2, you break down the second number, the number being multiplied.

For example in 6 x 7,

you break down the 7 into 5 + 2, and so

6 x 7	= 6 lots of 5	plus	6 lots of 2,
	= 6 x 5	+	6 x 2
	= 30	+	12
6 x 7	= 42		

It is easy to show the child how this works if you use coins. Use 2p plus 1p to represent 3p. Use 5p plus 1p to represent 6p. Use 5p plus 2p to represent 7p.

So, for example, 3 lots of 7p is made up from 3 lots of 5p plus 3 lots of 2p.

(5p) (2p) ☞ = 7p

(5p) (2p) ☞ = 7p

(5p) (2p) ☞ = 7p

☞ ☞

15p + 6p

3 lots of 7p = 7p + 7p + 7p = 21p
3 lots of 5p plus 3 lots of 2p
= 15p + 6p = 21p

For 3x

3 x 3 is   3 x 2   +   3 x 1   = 6 + 3 =   9

3 x 6 is   3 x 5   +   3 x 1   = 15 + 3 = 18

3 x 7 is   3 x 5   +   3 x 2   = 15 + 6 = 21

3 x 8 is   3 x 5   +   3 x 3   = 15 + 9 = 24

For 6x

6 x 6 is   6 x 5   +   6 x 1   = 30 + 6 = 36

6 x 7 is   6 x 5   +   6 x 2   = 30 + 12 = 42

6 x 8 is   6 x 5   +   6 x 3   = 30 + 18 = 48

More practice on 6x - 6 x 3

Set up 6 lots of 3. Use 2p and 1p coins to give a picture of what is happening. Remember to talk through each step.

2p	1p	= 3p
2p	1p	= 3p
2p	1p	= 3p
2p	1p	= 3p
2p	1p	= 3p
2p	1p	= 3p

12p + 6p → 18p

Set up one lot of 3p as a 2p and a 1p

Now line up altogether 6 lots of 2p and 1p

The 2p and 1p coins automatically separate the two parts of the sum.

The first part is   6 lots of 2   =   6 x 2   =   12

The second part is   6 lots of 1   =   6 x 1   =   6

Add the two parts   6 lots of 3   =   6 x 3   =   18

6 x 7

Again use the coins to give a picture of what is happening. Remember to talk through each step. Set up 6 lots of 7.

Set up one lot of 7p as a 5p and a 2p

Now line up altogether 6 lots of 5p and 2p

The 5p and the 2p coins automatically separate the sum into two parts.

The first part is	6 lots of 5	= 6 x 5	= 30
The second part is	6 lots of 2	= 6 x 2	= 12
Add the two parts	6 lots of 7	= 6 x 7	= 42

Now let the child try the remaining 6x facts.

Let him / her set up coins for;

      6 x 4   6 lots of 4

      6 x 6   6 lots of 6

      6 x 8   6 lots of 8

Practise 7x

Start by using coins, then try with just the numbers

7 x 7 is   7 x 5 + 7 x 2 = 35 + 14 = 49

7 x 8 is   7 x 5 + 7 x 3 = 35 + 21 = 56

5p 2p = 7p
5p 2p = 7p
5p 2p = 7p
5p 2p = 7p
5p 2p = 7p
5p 2p = 7p

30p + 12p → 42p

For 8x8

8 x 8 is   8 x 5 + 8 x 3 = 40 + 24 = 64.

All of these should be done, initially at least, with the child setting out the steps in coins and talking his / her way through each step of the strategy.

When these strategies are understood and mastered, then

0

to go!

THAT IS EVERY ONE OF

THE TIMES TABLE FACTS!

Section 2 of the book takes individual times table facts and provides a choice of strategies for each fact. Thus the child can use this section for review and revision.

SOME ODD FACTS

There are two unique facts in the table square. These are 3 x 4 and 7 x 8. Many people consider 7 x 8 to be the hardest fact to learn.

Show the child these patterns.   1234 and 5678.
Now insert some extra symbols

    12 = 3 x 4

    56 = 7 x 8

Squares

There is a pattern for square numbers and the numbers each side of them.

2 x 2 = 4	1 x 3 = 3
3 x 3 = 9	2 x 4 = 8
4 x 4 = 16	3 x 5 = 15
5 x 5 = 25	4 x 6 = 24
6 x 6 = 36	7 x 5 = 35
7 x 7 = 49	6 x 8 = 48
8 x 8 = 64	7 x 9 = 63
9 x 9 = 81	8 x 10 = 80

The answer to the square is always 1 bigger than the answer to the "each side' numbers.

Section

2

SECTION 2

In this section, each times table fact for 3x to 9x is dealt with separately, EXCEPT, each "reverse" fact, which is included in the heading. So, for example, if you want to look up 7 x 3, you may have to look for 3 x 7.

Most facts are answered by at least two methods. You must find the method which suits the child best. Reading and trying to understand the alternative methods may well help their general understanding, even though they eventually settle on one method. Be wary of confusing the child.

In this section, the 0x, 1x, 2x and 10x facts are not included. You must refer to the start of the book for these rather special cases.

Sometimes the presentation of similar examples is slightly different. This is done to extend experiences of mathematics. So you may see 5 x 6 or you may see 5 'lots of' 6 or you may see both. Remember that 'lots of' and 'x' are both used to mean times or multiply.

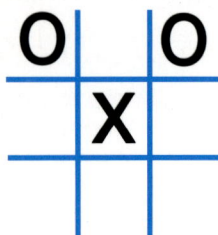

3 x 3 = 9

3 x 3 is a square.

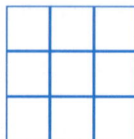

It is the familiar noughts and crosses shape.

Method 1

Using 3 x 3 as 3 'lots of' (2 + 1)

that is as 3 'lots of' 2 plus 3 'lots of' 1

So;

3 x 3	= 3 lots of 2	=	3 x 2	=	6
	+ 3 lots of 1	=	3 x 1	=	3 +
			3 x 3	=	**9**

Method 2

Using 3 x 3 as 2 lots of 3 plus 1 lot of 3

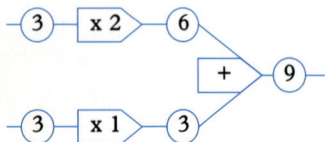

So;

3 x 3	= 2 lots of 3	=	2 x 3	=	6
	+ 1 lot of 3	=	1 x 3	=	3 +
			3 x 3	=	**9**

3 x 4 = 12   (which is also 4 x 3 = 12)

Method 1

3 x 4 = 12 is one of only two facts in the times table square where the answer and the question are in number order.

1234   becomes   12 = 3 x 4

Method 2

Turn 3 x 4 around to be 4 x 3. All 4x facts can be worked out by multiplying by 2 twice.

4 x 3   is done in two stages

Stage 1   2 x 3 = 6

Stage 2   2 x 6 = 12

Method 3

Use 1p coins to show that 3 x 4 is
2 lots of 4 plus 1 lot of 4

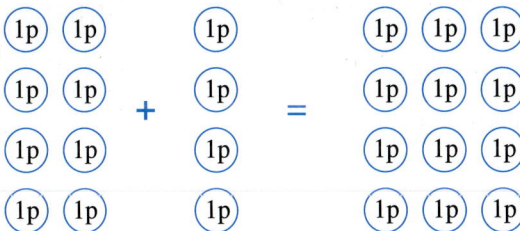

3 x 4   =   2 x 4   +   1 x 4

3 x 4   =   8   +   4   =   12

3 x 5 = 15   (which is also 5 x 3 = 15)

Method 1

Based on 5x facts being half the answer to the equivalent 10x fact.

$3 \times 5 = 3 \times 10 \div 2 = 15$

Method 2

Based on 3 x 5 being 3 'lots of' 5 which can be broken down into;

3 'lots of' 5 = 2 'lots of' 5 plus 1 'lot of' 5

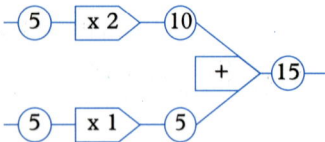

3 'lots of' 5 = 3 x 5 =

$$
\begin{array}{rl}
2 \times 5 & = 10 \\
1 \times 5 & = 5 + \\
\hline
3 \times 5 & = 15
\end{array}
$$

(Hint. Remember that the 5x table answers end in a 0 or a 5. If it is 5 x an odd number, there is a 5 at the end. If it is 5 x an even number, then there is a 0 at the end. This example was 3 x 5, 3 is an odd number, so the units digit must be a 5.)

3 x 6 = 18 (which is 6 x 3 = 18)

Method 1

Based on 6 x 3 as 6 'lots of' 3.

Which is; 5 'lots of' 3 plus 1 'lot of' 3

$$5 \times 3 \quad + \quad 1 \times 3$$

$$15 \quad + \quad 3$$

$$\mathbf{6 \times 3 = 18}$$

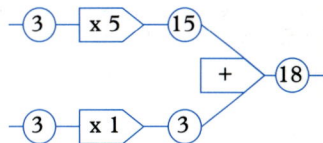

Method 2

Based on 3 x 6 as 3 'lots of' 6.

Which is; 2 'lots of' 6 plus 1 'lot of' 6

$$2 \times 6 \quad + \quad 1 \times 6$$

$$12 \quad + \quad 6$$

$$\mathbf{3 \times 6 \ = 18}$$

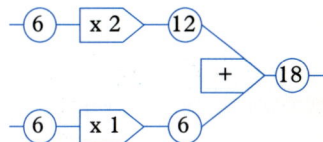

Method 3

Based on 3 x 6 as 3 x (5 + 1)

(Show this with 5p and 1p coins.)

$$3 \times 6 \ = \ 3 \times 5 \ + \ 3 \times 1 \ = \ 15 \ + \ 3 \ = \ 18$$

3 x 7 = 21   (which is also 7 x 3 = 21)

Method 1

Based on 7 x 3 and using 5 + 2 for 7.

So 7 'lots of' 3 is made up from 5 'lots of' 3 plus 2 'lots of' 3.

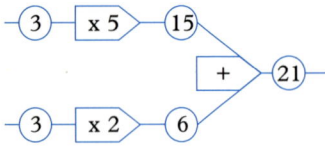

Multiply 5 x 3   = 15

Multiply 2 x 3   = 6

Add the two parts together

15 + 6 = 21   7 x 3 = 21

This shows how 7x can be calculated by breaking it down into 5x plus 2x.

Method 2

Based on 3 x 7 as 3 'lots of' 7 and using '2 lots of' 7 plus '1 lot' of 7.

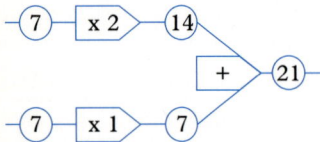

2 'lots of' 7   = 2 x 7 = 14

1 'lot of' 7   = 1 x 7 = 7

Add the two parts together

14 + 7 = 21

3 x 7 = 21

3 x 8 = 24   (which is also 8 x 3 = 24)

Method 1

Based on 3 x 8 as 3 'lots of' 8 which is 2 'lots of' 8

plus 1 'lot of' 8

3 x 8	=	2 lots of 8	+	1 lots of 8	
	=	2 x 8	+	1 x 8	
	=	16	+	8	= 24

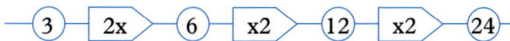

(Hint.... add 8 to 16 in two stages.....

$$16 + 4 = 20 \quad 20 + 4 = 24).$$

Method 2

Based on 8 being 2 x 2 x 2.

So to multiply 3 by 8, you multiply 3 by 2 by 2 by 2;

3 x 8   3 x 2 = 6   6 x 2 = 12   12 x 2 = 24

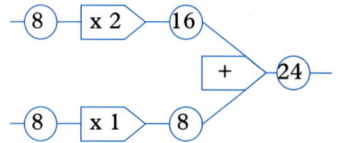

3 x 9 = 27   (which is 9 x 3 = 27)

Method 1

Based on 3 x 9 being 3 'lots of' 9;

which is 2 'lots of' 9 plus 1 'lot of' 9.

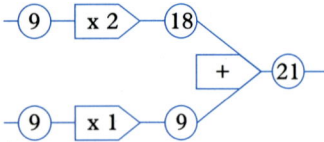

$2 \times 9 = 18$

$1 \times 9 = 9 +$

$3 \times 9 = 27$

Method 2

Based on 3 'lots of' 9 being 3 'lots of' 10 less than 3 'lots of' 1.

That is;   3 x 9 = 3 x 10 - 3 x 1

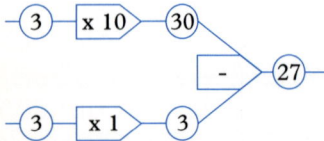

$3 \times 10 = 30$

$3 \times 1 = 3 -$

$3 \times 9 = 27$

(Hint..... The individual digits or numbers which make up the answers to all 9x tables facts add up to 9. In this case the answer was 27 and $2 + 7 = 9$. This acts as a check on the answer.)

4 x 4 = 16

This is a square.

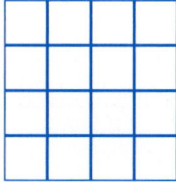

The square numbers follow a sequence.

1   4   9   16   25   36   49   64   81   100

The difference (the amount you must add on between each successive number) goes up as the odd numbers go up.

$$4 - 1 = 3$$
$$9 - 4 = 5$$
$$16 - 9 = 7$$
$$25 - 16 = 9, \quad \text{etc.}$$

So 4 x 4 is 7 more than 3 x 3. The answer is 9 + 7 = 16.

4 x 4 can also be worked out by using two stages;
4 x 2 = 8   and   8 x 2 = 16.

To work out 4 x 4 you multiply 4 by 2 two times.

4 x 5 = 20   (which is also 5 x 4 = 20)

Method 1

To multiply by 4 it is often easier to multiply by 2 twice (since 4 is 2 x 2).

4 x 5 can be worked out in two stages.

2 x 5 = 10

2 x 10 = 20   so 4 x 5 = 20.

Method 2

To multiply by 5 it is often easier to multiply by 10 then divide by 2. 5 x 4 can be worked out in two stages.

10 x 4 = 40

40 ÷ 2 = 20   so 5 x 4 = 20.

(Check. Even numbers times 5 give an answer that ends in 0; 4 is an even number, 20 ends in 0).

4 x 6 = 24   (which is also 6 x 4 = 24)

Method 1

It is easy to multiply by 2 twice. This is the same as multiplying by 4.

Sometimes two easy stages are better than one hard stage.

2 x 6 = 12

2 x 12 = 24   so 4 x 6 = 24.

(Extra information. 3 x 8 also equals 24. When you break down numbers it can help to show why this is so. 4 x 6 breaks down to 2 x 2 x 6. This can be further broken down by using the fact that 6 = 2 x 3. So 4 x 6 becomes;

2 x 2 x 2 x 3.

2 x 2 x 2 = 8   and 8 x 3 = 24

so 4 x 6 and 8 x 3 are the same).

Method 2

6 x 4 can be 6 'lots of' 4,

and 6 'lots of' 4 can be broken down into 5 'lots of' 4 plus 1 'lot of' 4. The two parts are then added together again to give 6 x 4.

5 'lots of' 4 = 5 x 4 = 20

1 'lot of' 4 = 1 x 4 = 4 +
_____

6 'lots of' 4 = 6 x 4 = 24

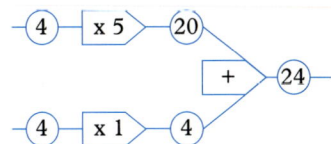

73

4 x 7 = 28 (which is also 7 x 4 = 28)

Method 1

Based on 4 x 7 and using 2 x 2 as 4. To multiply by 4 it can be easier to multiply by 2 twice.

7 — $\times 2$ — 14 — $\times 2$ — 28

| Multiply | 2 x 7 | = | 14 |
| Multiply | 2 x 14 | = | 28 |

4x can be calculated by doubling and then doubling again.

Method 2

Based on 7 x 4 and using 7 as 5 + 2

So 7 'lots of' is made up of 5 'lots of' plus 2 'lots of'

4 — $\times 5$ — 20

4 — $\times 2$ — 8

$+$ — 28

| Multiply | 5 x 4 | (which is 5 'lots of' 4) | = | 20 |
| Multiply | 2 x 4 | (which is 2 'lots of' 4) | = | 8 |

Add the two parts together

20 + 8 = 28   so   7 x 4 = 28

7 x 4 can be calculated by working out 5 x 4 and then adding on 2 x 4.

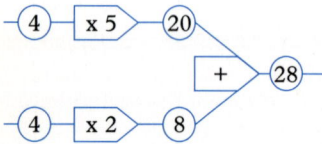

4 x 8 = 32   (which is also 8 x 4 = 32)

32 is a part of the sequence of numbers you get when you start with 1 and keep doubling.

1   2   4   8   16   32   64   128   256   512   1024

4 is 2 x 2   and   8 is 2 x 2 x 2   so 4 x 8 and 8 x 4 are made up only of 2x.

Both are made up of 2 x 2 x 2 x 2 x 2.

Method 1

4 is 2 x 2, so you can multiply by 4 by multiplying by 2 twice.

2 x 8 = 16

2 x 16 = 32   so 4 x 8 = 32.

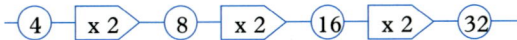

Method 2

8 x 4 is 8 'lots of' 4.

8 'lots of' 4 can be broken down into 5 'lots of' 4 plus 3 'lots of' 4.

5 'lots of' 4 = 5 x 4 = 20

3 'lots of' 4 = 3 x 4 = 12

Add these two parts together to give 8 'lots of' 4

= 8 x 4 = 32

1 + 1 = 2

2 + 2 = 4

4 + 4 = 8

8 + 8 = 16

16 + 16 = 32

32 + 32 = 64

64 + 64 = 128

128 + 128 = 256

256 + 256 = 512

512 + 512 = 1024

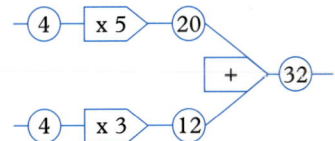

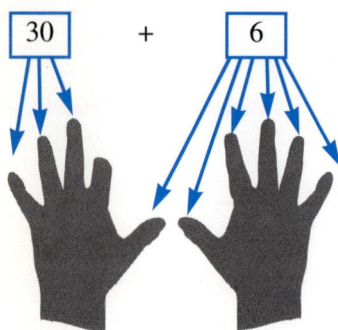

| 30 | + | 6 |

This finger method is useful for the full 9x table. By always dropping one finger you end up with

x2 = 1....8
x3 = 2....7
x4 = 3....6
x5 = 4....5
x6 = 5....4
x7 = 6....3
x8 = 7....2
x9 = 8....1
x10 = 9....0

And of course if you add the two numbers together they must always equal 9!

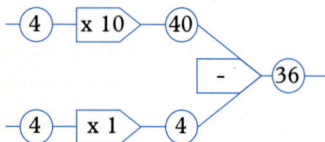

4 x 9 = 36   (which is also 9 x 4 = 36)

Method 1

The finger method. Place all ten fingers on a table edge (remember that in maths, thumbs count as fingers). Tuck the fourth (number 4) finger from the left under the table. To the left of this tucked away finger are 3 fingers, and to the right of it are 6 fingers. This gives you the answer.....36.

Method 2

Using 2 x 2 = 4. Instead of multiplying by 4 in one stage, you multiply in two stages, both stages by 2x.

Stage 1.   2 x 9 = 18

Stage 2.   2 x 18 = 36

Method 3

Using 9 x 4 for 4 x 9 and using 10 'lots of' 4 less 1 'lot of' 4.

Stage 1.   10 x 4 = 40

Stage 2.   1 x 4 = 4

Stage 3.   40 - 4 = 36

(Check: 36.......3 + 6 = 9......all 9x answers add to 9).

5 x 5 = 25

5 x 5 is a SQUARE. 5 is multiplied by itself.

25 is one quarter of 100. Since 100 is the answer to 10 x 10 and 5 is half of 10, then 5 x 5 should give an area that is 1/4 of 100.

Method

5 is half of 10.

Any 5x table fact has an answer that is half of the equivalent 10x table fact.

Stage 1.   10 x 5 = 50

Stage 2.   50 ÷ 2 = 25

5 x 5 = 25

5 x 6 = 30   (which is also 6 x 5 = 30)

Method 1

Based on 5 'lots of' a number giving an answer which is half of the 10 'lots of' the same number.

So 5 'lots of' 6 is half of 10 'lots of' 6.

10 x 6 = 60

Half of 60 = 30   5 x 6 = 30

(An even number times 5 always has a 0 as its units digit).

Method 2

Based on 6 x 5 as 6 'lots of' 5.

And on 6 'lots of' 5 being equal to 5 'lots of' 5 plus 1 'lot of' 5.

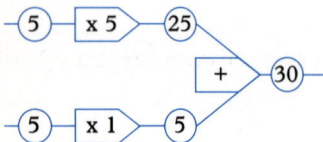

5 'lots of' 5 = 5 x 5 = 25

1 'lot of' 5 = 1 x 5 =   5

Add the two parts together to make 6 'lots of' 5,

6 x 5 = 25 + 5 = 30.

6 x 5 = 30

5 x 7 = 35   (which is also 7 x 5 = 35)

Method 1

Based on 7 x 5 as 7 'lots of' 5

and breaking down 7 'lots of' 5 into

5 'lots of' 5 plus 2 'lots of' 5

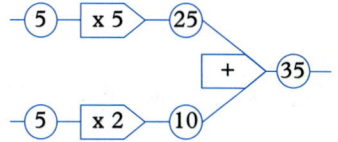

Multiply   5 x 5   =   25   (5 lots of 5)

(remember that 5x facts are half of the same 10x fact,

5 x 10 = 50)

Multiply   2 x 5   =   10   (2 lots of 50

Add the two parts together

25 + 10 = 35   **7 x 5 = 35**

Method 2

Based on 7 x 5 as 5 x 7

and on the 5x facts being half the value of 10x facts.

10 x 7 = 70

70 ÷ 2 = 35   **5 x 7 = 35**

(Remember an odd number x 5 gives an answer which

ends in 5 and 7 is an odd number).

5 x 8 = 40   (which is also 8 x 5 = 40)

Method

The easiest method is to use the fact that 5x table answers are half the value of the equivalent 10x table answer.

The answer is obtained by two stages. First multiply 8 by 10 to get 80, then divide 80 by 2 to obtain 40.

Stage 1.   10 x 8 = 80

Stage 2.   80 ÷ 2 = 40

(Remember an even number times 5 gives an answer which ends in 0.)

$8 - \boxed{\text{x } 10} - 80 - \boxed{÷ 2} - 40$

5 x 9 = 45   (which is also 9 x 5 = 45)

Method 1

The "finger" method for 9x. Place all ten fingers on the edge of a table. Tuck in finger number 5 (counting from the left, this is the thumb on your left hand).

The fingers to the left of the tucked in finger give the tens digit (4) and the fingers to the right of the tucked in finger give the units digit (5). Answer is 45.

Method 2

Based on 5 x 9 being half of 10 x 9.

Multiply 10 x 9   =   90

(remember the pattern for 10x)

Divide 90 by 2   =   45

(to 'half' a number, divide it by 2)

Method 3

Based on 9 x 5, and based on 9 being 1 less than 10, and based on each 9 in a multiplication being 1 less than each 10.

Multiply 10 x 5   =   50

5 lots of 9 are 5 lots of 1 less than 5 lots of 10

(remember the pattern)

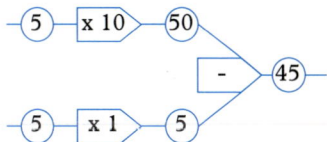

So, subtract 5 from 50   50 - 5 = 45

Remember the digits in the answers to all 9x sums add up to 9.   4 + 5 = 9

81

6 x 6 = 36

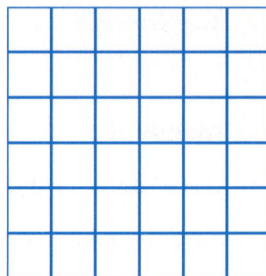

6 x 6 is a square. The two
multiplying numbers are the
same as each other.

6 is multiplied by itself, 6.

Method 1

Based on 6 x 6 as 6 'lots of' 6.

And on 6 as '5 plus 1', and working with the first 6.

So 6 'lots of' 6 can be seen as 5 'lots of' 6 plus 1 'lot
of' 6.

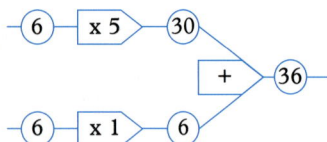

6 x 6 = (5 x 6) + (1 x 6)

          (5 lots of 6 and one more 6)

     =   30 + 6   =   36

6 x 6 = 36

(Remember that 5 x 6 is half of 10 x 6).

Method 2

Based on 6 as '5 plus 1', and working with the second 6.

So 6 'lots of' 6 can be seen as 6 'lots of' 5 plus 6 'lots
of' 1.

6 x 6 = (6 x 5) + (6 x 1)

     =   30 + 6   =   36

6 x 6 = 36

6 x 7 = 42   (which is also 7 x 6 = 42)

Method 1

Based on 7 x 6 and using 7 as 5 + 2

So 7 'lots of' 6 is made up from 5 'lots of' 6 plus 2 'lots of' 6.

Multiply 5 x 6   =   30

    (remember 5 x 6 is half of 10 x 6)

Multiply 2 x 6   =   12

Add the two parts together   30 + 12 = 42

7 x 6 = 36

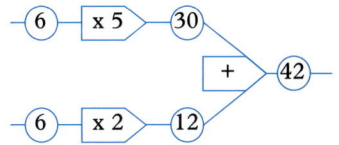

Method 2

Based on 6 x 7 and using 6 as 5 + 1.

So 6 'lots of' 7 is made up from 5 'lots of' 7 plus 1 'lot of' 7.

Multiply 5 x 7   =   35

(remember 5 x 7 is half of 10 x 7)

Multiply 1 x 7   =   7

Add the two parts together

35 + 7 = 42   (remember you can add 7 in two steps,

        35 + 5 = 40   40 + 2 = 42)

6 x 7 = 36

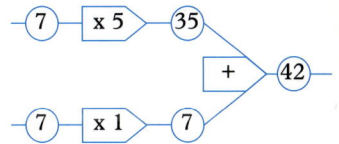

6 x 8 = 48   (which is also 8 x 6 = 48)

Method 1

Using 6 x 8 as 6 'lots of' 8 and breaking down 6 'lots of' 8 to 5 'lots of' 8 plus 1 'lot of' 8.

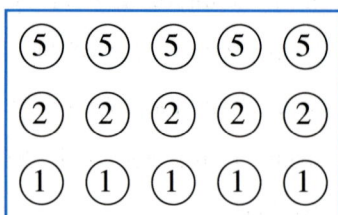

5 'lots of' 8   1 'lots of' 8

Stage 1.   5 'lots of' 8 = 5 x 8 = 40

Stage 2.   1 'lot of' 8 = 1 x 8 = 8

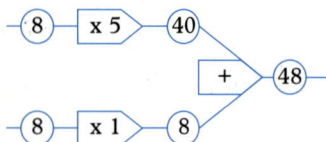

Stage 3.   Add the two parts together.

5 'lots of' 8 + 1 'lot of' 8   =   40 + 8 = 48

6 x 8 = 48

Method 2

Using 8 x 6, and multiplying by 8 in two stages. First by 4, then by 2.

Stage 1.   4 x 6 = 24

Stage 2.   2 x 24 = 48

8 x 6 = 48

6 x 9 = 54   (which is also 9 x 6 = 54)

Method 1

As 9 x 6. Use the finger method for the 9x table. Place all ten fingers on the edge of a table. Tuck the sixth finger from the left (this will be the thumb on your right hand) under the table. The 5 fingers to the left of the tucked in finger give the tens digit and the 4 fingers to the right of the tucked in finger give the units digit The answer is 54.   **9 x 6 = 54**

Method 2

Use 6 x 9. Remember that 9x facts can be worked out from 10x facts.

6 nines are 6 ones less than 6 tens.

6 x 9 = 6 x 10 - 6 x 1 = 60 - 6 = 54

6 x 9 = 54

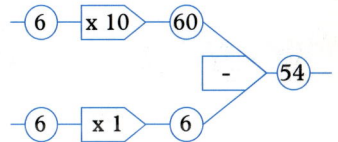

Method 3

Using 6 x 9 as 6 'lots of' 9 which can be split into 5 'lots of' 9   plus   1 'lot of' 9. This makes three stages.

Stage 1.   5 x 9   = 45

Stage 2.   1 x 9   = 9

Stage 3.   45 + 9 = 54

(Check if the answer follows the 9x rule....... 5 + 4 = 9).

6 x 9 = 54

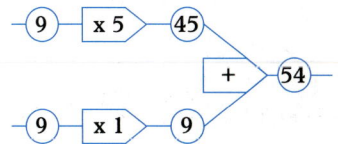

7x7 = 49

7 x 7 is a square. 7 is multiplied by itself. We used an area as one picture for these times table facts and the area picture for 7 x 7 is a square.

Method 1

Using 7 x 7 as 7 'lots of' 7

And 7 'lots of' 7 as 5 'lots of' 7 plus 2 'lots of' 7. (This is breaking up the first 7).

So;

7 x 7 = 5 'lots of' 7 + 2 'lots of' 7

7 x 7 = 5 x 7 + 2 x 7

7 x 7 = 35 + 14 = 49

7 x 7 = 49

Method 2

Using 7 x 7 as 7 'lots of' (5 plus 2). (This is breaking up the second 7).

So;

7 x 7 = 7 x (5 + 2)

7 x 7 = 7 x 5 + 7 x 2

7 x 7 = 35 + 14 = 49 **7 x 7 = 49**

Note

The biggest answer in this times table book is 100, which is the answer for 10 x 10. The answer to 7 x 7 is almost half of 100. It is 1 less than 50.

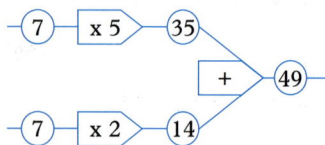

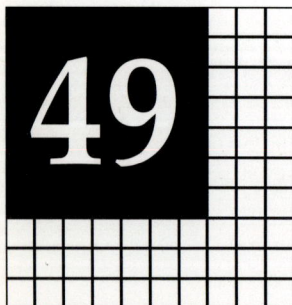

7 x 8 = 56   (which is also 8 x 7 = 56)

Note. If you reverse the normal order of presentation to be 56 = 7 x 8, this fact has its four numbers in order of counting, 5 6 7 8.

Method 1

Using 7 x 8 as 7 'lots of' 8 and breaking 7 'lots of' 8 into 5 'lots of' 8 plus 2 'lots of' 8. The two parts are then combined to give the answer. Work through this with coins.

7 x 8 = 7 'lots of' 8 = 5 'lots of' 8   +   2 'lots of' 8

7 x 8   =   5 x 8   +   2 x 8

7 x 8   =   40   +   16

                =   56

7 x 8 = 56

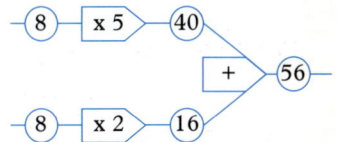

Method 2

Using 8 x 7 and then, in a strategy similar to the one used for the 9x table, work back from 10 x 7.

8 x 7 is 8 'lots of' 7.

8 'lots of '7 is 10 'lots of' 7 less 2 'lots of' 7.

8 x 7 = 10 x 7 -   2 x 7

8 x 7 =   70   -   14   =   56

(It may be easier to take 14 away in two steps, first take off 10 to reach 60, then 4 to give 56).

8 x 7 = 56

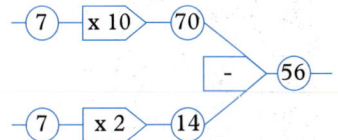

$7 \times 9 = 63$ (which is also $9 \times 7 = 63$)

Method 1

Use 9 x 7 and the finger method for 9x facts.

Place all ten fingers on the edge of a table. Tuck the seventh (counting from the left) finger under the table. This should be the index finger of your right hand.

The number of fingers to the left of this tucked in finger gives the tens digit......6.

The number of fingers to the right of the tucked in finger gives the units digit......3.

The answer is 63.   **9 x 7 = 63.**

Method 2

Using 9 x 7 and working via the estimate 10 x 7.

The pattern for 9x facts is;

7 'lots of' 9 is 7 'lots of' 10 less than 7 'lots of' 1.

Thus 7 x 9 is 7 less than 7 x 10.

Step 1.   7 x 10 = 70

Step 2.   70 - 7 = 63

Check.   6 + 3 = 9   (Remember, the digits of the 9x table answers all add to make 9)

9 x 7 = 63

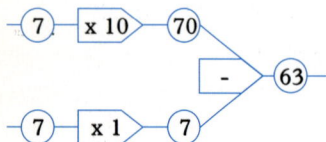

8 x 8 = 64

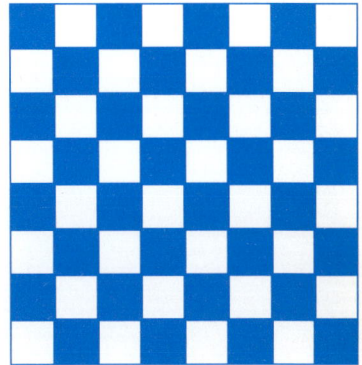

Method 1

8 x 8 is a SQUARE number. 8 is multiplied by itself.

Changing the order of multiplying the two numbers looks no different as both numbers are the same.

Each of the two 8's is made up of 2 x 2 x 2.

Together, 8 x 8 is 2 x 2 x 2 x 2 x 2 x 2.

As each 2 is used to multiply, the number grows; 2, 4, 8, 16, 32, 64. This sequence of repeated doubling is quite easy to master and relate to **8 x 8.**

Method 2

Each 8 is worth 1 'lot of' 10 less 1 'lot of '2.

(1 x 8 = 1 x 10 - 1 x 2).

So 8 'lots of' 8 are 8 'lots of' 10 less 8 'lots of' 2.

8 x 8 = 8 x 10 - 8 x 2

       = 80   -   16   =   64

8 x 8 = 64

Method 3

Using 8 'lots of' 8 as 5 'lots of' 8 plus 3 'lots of 8'

8 x 8 = (5 x 8) + (3 x 8)

      =   40   +   24   =   64   **8 x 8 = 64**

(The answer must be an even number because you are multiplying two even numbers together)

The square 8 x 8 is of course the chessboard. The inventor of chess was offered any reward he would like for its invention. He said he would like a grain of rice on the first square, 2 grains on the second square, 4 on the third, 8 on the fourth, and so on, doubling every square.

If you work it out you will discover that the whole world has never produced this much rice in all history!

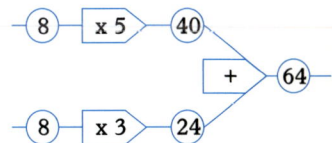

8 x 9 = 72   (which is also 9 x 8 = 72)

Method 1

Using 9 x 8 and the finger method for 9x table facts.

Place all ten fingers on a table edge. Tuck down the eighth finger counting from the left (this is the middle finger on your right hand).

Count the fingers to the left of the tucked in finger. There are seven. 7 is the tens digit of your answer. Now count the fingers to the right of the tucked down finger. There are two. 2 is the units digit of your answer.

9 x 8 = 72

Method 2

Using 8 x 9 and working via an estimate of 8 x 10 and the pattern for this method.

The pattern is;

8 'lots of' 9 are 8 'lots of' 10 less 8 'lots of' 1.

which can be said as; 8 nines are 8 less than 8 tens.

8 x 9 = 8 x 10   -   8 x 1

8 x 9 =   80   -   8

8 x 9 =   72

(Check 7 + 2 = 9)

8 x 9 = 72

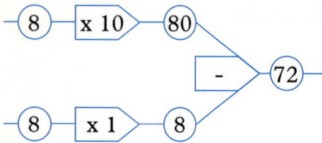

Conclusion

So what do we do now *practise, practise, practise.*

That is the only way to learn these facts. Whether it is in the supermarket, or the newsagent, or counting out pocket money, always practise the methods.

Do not forget that these techniques may also be applied to bigger numbers. The same rules apply!

You can easily work out for youself a strategy for

11x, (use 10x plus 1x)

12x, (use 10x plus 2x)

15x, (use 10x plus half of 10x)

20x, (use 2x then 10x)

50x, (use 5x then 10x)

or even 99x (use 100x minus 1x).

Table Squares

	0	1	2	3	4	5	6	7	8	9	10
0	0	0	0	0	0	0	0	0	0	0	0
1	0	1	2	3	4	5	6	7	8	9	10
2	0	2	4	6	8	10	12	14	16	18	20
3	0	3	6	9	12	15	18	21	24	27	30
4	0	4	8	12	16	20	24	28	32	36	40
5	0	5	10	15	20	25	30	35	40	45	50
6	0	6	12	18	24	30	36	42	48	54	60
7	0	7	14	21	28	35	42	49	56	63	70
8	0	8	16	24	32	40	48	56	64	72	80
9	0	9	18	27	36	45	54	63	72	81	90
10	0	10	20	30	40	50	60	70	80	90	100

Calculating strips

Please cut out, or photocopy this page and use the strips
as shown in the book.

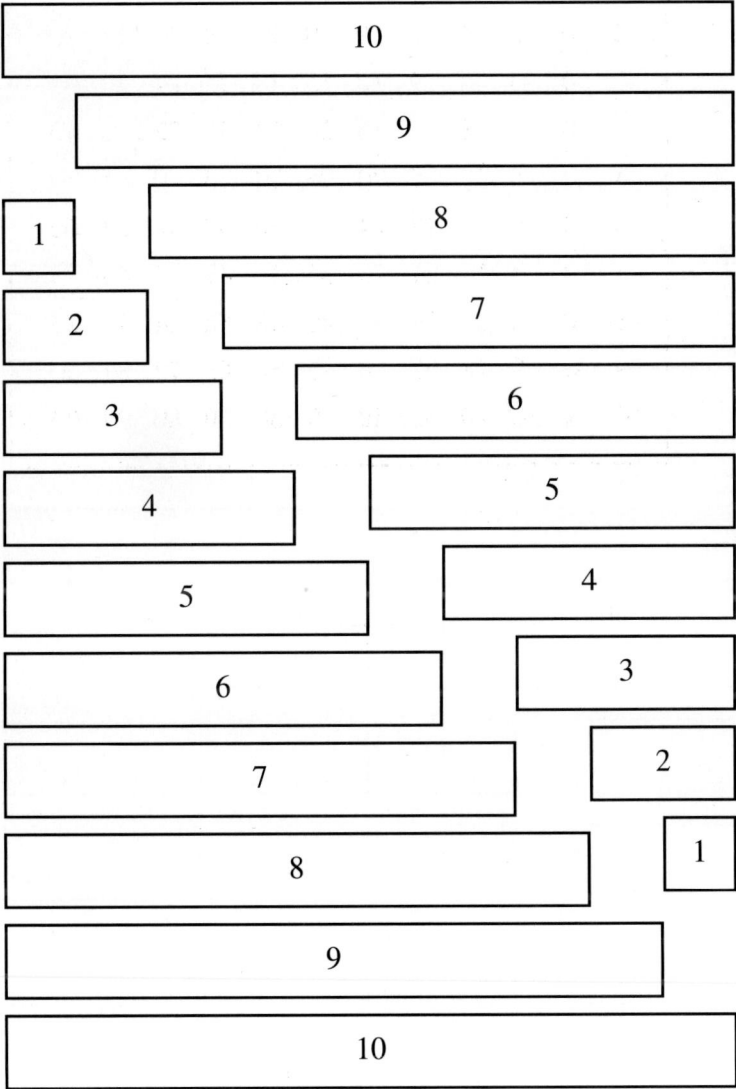

10	

9	

1	8

2	7

3	6

4	5

5	4

6	3

7	2

8	1

9	

10	

More strips

Please photocopy this page and use it with those over on page 93.

9

9

9

9

10

10

10

10

2	1	5
2	1	5
2	1	5
2	1	5